BUILDING BLOCKS OF LIFE

A Scribner Portfolio in Natural History

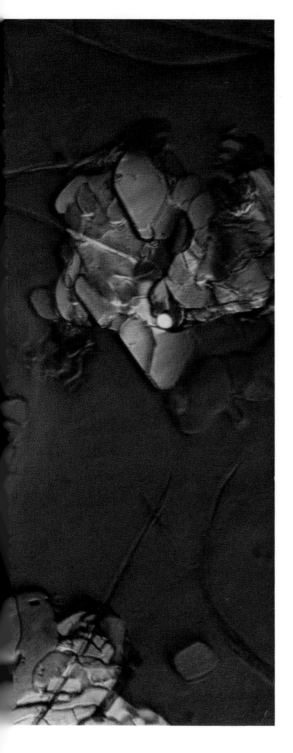

BUILDING BLOCKS OF LIFE

Proteins, Vitamins, and Hormones Seen Through the Microscope

Photographs and text by
ROMAN VISHNIAC

Charles Scribner's Sons • New York

Copyright © 1971 Roman Vishniac

Title-page illustration:

Deoxyribonucleic acid (DNA)

This book published simultaneously in the United States of America and in Canada—Copyright under the Berne Convention

All rights reserved. No part of this book may be reproduced in any form without the permissision of Charles Scribner's Sons.

A–8.71 [CTZ]

Printed in the United States of America
SBN 684–12381–9
Library of Congress Catalog
Card Number 76-143959

BUILDING BLOCKS OF LIFE

The basic substances of life, the bricks and mortar of vitality, include not only the nucleic acids such as DNA but many other substances such as vitamins, hormones, and enzymes. All of these substances are essential to the processes of life.

Life consists of a variety of complex and perplexing phenomena. To describe the processes involved in fertilization, enzyme action, susceptibility and immunity, or growth and evolution is only to deal with separate aspects of the basic question—what is life? After learning the "hows" man demands knowledge of "why." The most baffling questions still remain unanswered. The horizons of the human mind expand more rapidly than the information that can be gained from nature. For every question answered a hundred new ones arise.

The pictures in this book represent a selection of basic substances, but they are not portraits for use in medical textbooks. Rather, they demonstrate the startling similarities between living cells seen through an interference microscope and the canvases of an abstract painter.

To watch colors change through polarization, changes of phase, and differentiation by interference is an unforgettable experience. Interference—the interaction of two light waves crossing at a glancing angle—changes the form and color of the images. Optical systems using interference have been developed since 1950 and provide increased visibility compared with the phase-contrast instruments in use since the 1940s. The combinations of colors produced are artistic, not natural. Contrasts are sharpened and the microscopist who is artistically minded is able to create compositions of infinite variety.

Proteins

Protein came to the attention of the general public in the 1840s. On July 10, 1838, the Swedish scientist Jöns Jakob Berzelius (1779–1848), master of chemistry and developer of the modern system of chemical symbols and formulas, wrote a letter to his younger colleague, the Dutch chemist Gerard Johann Mulder (1802-1880), in which he suggested a new name for the material they had been discussing in their previous correspondence. The newly coined term, "protein," was derived from the Greek *proteios* (first), indicating the primary or basic nature of the material. Mulder liked the name and began to use it immediately. In his widely read treatise, *Proeve eener algemeena physiologische Scherkunde* (Toward a General Physiological Chemistry, 1843–1850), he said of protein: "It is the most important of the known components of living matter. Without it life would be impossible."

Scientists had studied these substances for several decades before giving them their present name. Much later, in the twentieth century, proteins (1)* were found to be highly complex nitrogenous substances with molecular weights ranging from 10,000 to many millions. In the human body there are thousands of different proteins, each having a special chemical structure and performing a specific task. When heated in either acidic or alkaline solutions, proteins decompose and break up into amino acids —a process known as hydrolysis. The simplest of the amino acids, glycine (2),

*Numbers in parentheses refer to the illustrations.

1 Silk fibroin, protein

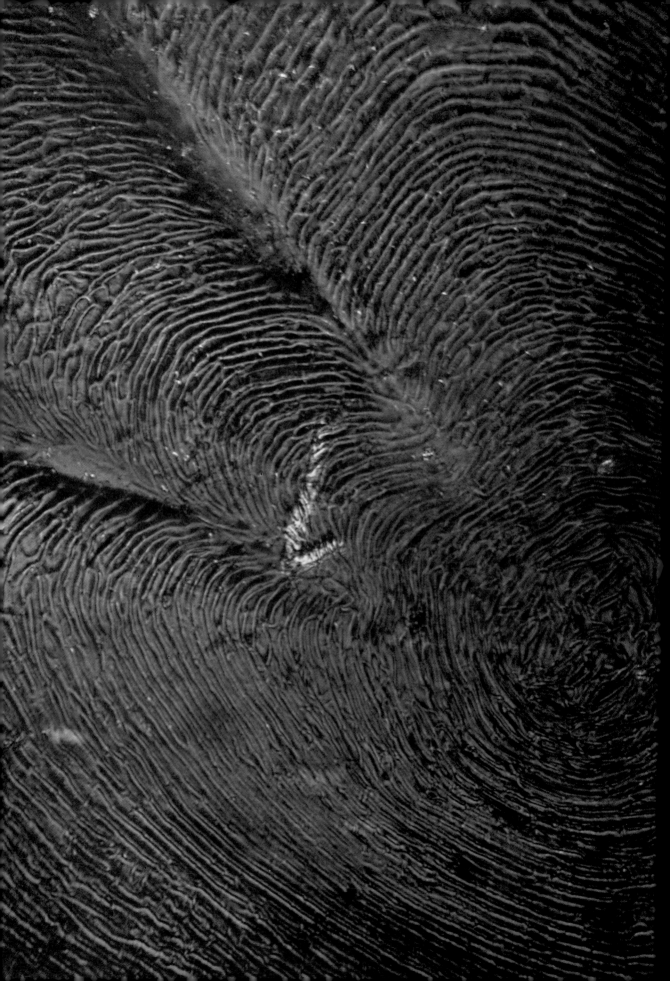

was isolated in 1820 by the French scientist Henri Braconnet (1781–1855); its name was coined by Berzelius.

Of the twenty-three recognized amino acids of proteins, nine must be obtained by man from protein foods; among these are histidine (3), tryptophan (4), lysine (5), and leucine (6). The remaining acids are synthesized in the body. The ability of the human body to manufacture amino acids from inorganic components is remarkably limited. Even microorganisms have a greater capacity for amino-acid manufacture than does man and for that reason are more capable of adapting quickly to changes in the environment. Not all protein foods have equal value; they are classed as good or poor, depending on whether or not they contain all nine amino acids required in man's diet. An example of a good protein is casein, the chief protein of milk; a poor one is gelatin, manufactured by boiling bones.

2 Glycine, amino acid *(left)*

3 Histidine, amino acid, from salmon sperm

4 Tryptophan, amino acid *(overleaf)*

5 Lysine, amino acid

6 Leucine, amino acid

In a classic monograph, *Untersuchungen über Amino säuren, Polypeptide und Proteine* (Investigations on Amino Acids, Polypeptides, and Proteins), the German chemist Emil Fischer (1852–1919) showed that proteins are composed of amino acids linked by elimination of water from their separate molecules. He called these linkages polypeptide chains. Beginning with the linking of two molecules, the process can continue until several long chains are formed. By continued hydrolysis, these chains can be broken up and the number and arrangement of amino acids in a protein molecule can be determined.

The number of possible combinations of amino acids has been calculated to be two billion billion; amino acids can therefore produce an endless diversity of proteins. Many are important components of animal food.

The cells which in combination form the tissues of the human body consist largely of proteins. In addition, the body contains bones—created by bone-making cells and consisting of a combination of inorganic materials with the protein collagen—and fluids, in which other cells float. The red cells in the blood are 35 per cent hemoglobin (7), a protein containing iron and having the property of combining reversibly with oxygen. This enables the blood to combine with oxygen in the lungs and carry it through the blood vessels to the tissues, where it is liberated to oxidize food molecules. At the same time, the blood absorbs the carbon dioxide produced in the tissues by oxidation and carries it back to the lungs to be exhaled.

As long as hemoglobin retains the power of combining reversibly with oxygen, it appears in its "native" form. A protein which has lost its characteristic properties is said to be "denatured." Hemoglobin, like most proteins, can be denatured by heating to about 65°C.

When an egg is cooked, the soluble protein ovalbumin is denatured into an insoluble hard coagulum. The denaturation process is thought to involve the uncoiling of the polypeptide chains. In a cooked egg, as with any heated protein, the uncoiled poylpeptide chains become entangled and cannot be separated. The denatured protein is therefore insoluble.

In the human body, proteins in food are broken down into amino acids which pass into the blood stream. Under laboratory conditions, hydrolysis is accomplished by subjecting proteins to strong acids and to boiling over a long period at high temperatures, but at the temperature of the body proteins cannot be broken down without special catalytic substances found in living organisms. These substances, called enzymes, are proteins which guide many aspects of life processes. Like other catalysts, they have the property of accelerating chemical reactions without themselves undergoing change. There are estimated to be 36,000 different enzymes in the human body. Each serves as a catalyst for a specific chemical reaction, most of which are beneficial for the organism.

7 Hemoglobin crystals, protein

Enzymes and digestion

Until the eighteenth century, vaguely defined terms such as "coction" or "fermentation" were generally used to describe the changes which food underwent in the stomach. During that century, however, several investigators studied the action of the juices found in the stomach and intestines. One of the most important of these investigators was the Frenchman René Antoine Réaumur (1683–1757), inventor of the temperature scale which bears his name. His experiments are described in his memoir "Sur la digestion des oiseaux" (On the Digestion of Birds, 1752). In one of his major experiments, he induced a buzzard to swallow a sponge attached to a thin thread. After some time the sponge was recovered, drenched with gastric juice from the stomach (8). This juice was applied experimentally to various kinds of edible materials. Next he let animals swallow meat in wire-mesh containers or in perforated capsules, and retrieved these by attached strings. By these experiments Réaumur greatly clarified the action of gastric juice.

An important contribution was made seventy years later in the United States. In 1822 Alexis St. Martin, a French-Canadian visitor at the American Fur Exhibition on Mackinac Island in Lake Huron, received by accident a terrible gunshot wound. William Beaumont (1785–1853), a young apprentice to a surgeon in the U.S. Army Medical Corps, tried to save St. Martin's life by performing several surgical operations. Despite the severity of the wound, the patient survived. When after ten months the local hospital refused to keep the young man as a

8 Gastric juice of chicken

patient any longer, Dr. Beaumont took St. Martin into his home.

After two years of care, St. Martin had been restored to health, even though a narrow gap remained in a normally continuous area of abdominal tissue. This narrow hole was covered only by a thin membrane, which could easily be pushed aside. St. Martin, in other words, was left as a result of his wound with a permanent gastric fistula, a window into his stomach. This fistula remained until his death at the age of eighty-three. Dr. Beaumont realized that he was thereby provided with a rare opportunity to study directly the action of human gastric juice.

"When he lies on the opposite side," Beaumont wrote in his notebooks, "I can look directly into the cavity of the stomach and see the process of digestion. I have suspended flesh and other substances into the perforation. This case affords an excellent opportunity for experimenting."

As Réaumur did with birds, Beaumont inserted different substances on a thread into St. Martin's stomach and extracted gastric juice (9) for experimentation. As far as the human body is concerned, this "open window" into St. Martin's stomach has retained its special place in the history of medicine. Not until the twentieth century did Pavlov fully exploit the advantages of gastric fistulas by producing them experimentally in dogs.

Meanwhile, however, several investigators pressed the search for the active substances in the digestive juices. By 1838 the German microscopist Theodor Schwann (1810-1882)) had isolated and named pepsin (10), the main enzyme of

9 Human gastric juice

10 Pepsin, enzyme

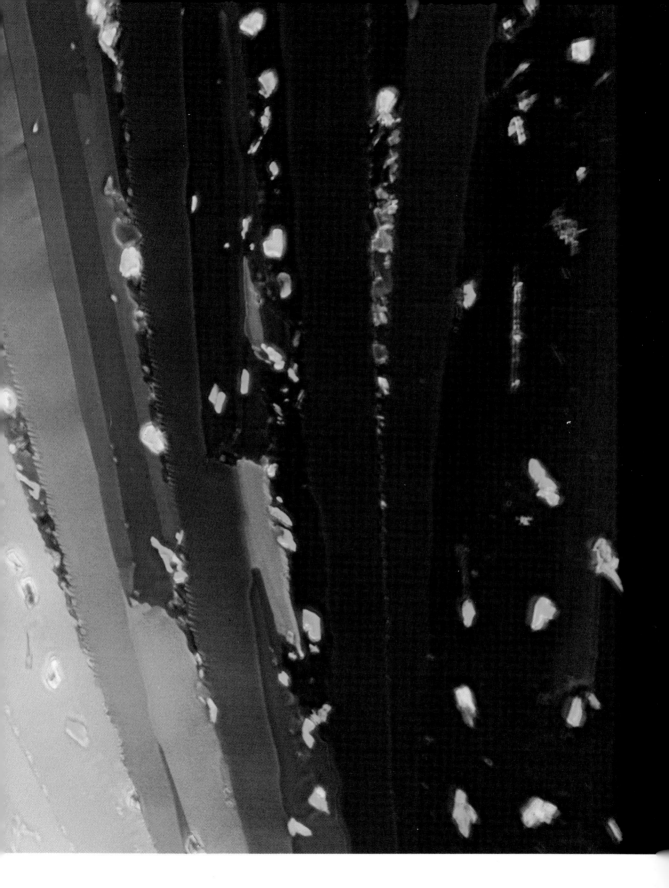

the stomach. Pepsin, together with the other enzymes in the digestive juices, breaks food down into amino acids, which are then able to pass through the walls of the gastrointestinal tract into the bloodstream. Food elements in the form of amino acids are thus carried through the blood vessels to the tissues. They are the building blocks of the body's proteins. In hospitals, food is often supplied to patients by injecting a solution of amino acids directly into the bloodstream.

Enzymes and fermentation

As early as 1837, Schwann—who was also the chief architect of the theory that animals and plants consist of cells—argued that a living microorganism, yeast, was responsible for alcoholic fermentation. Although the same conclusion was reached almost simultaneously by the French physicist Baron Charles Cagniard-Latour (1777–1859), and by the German phycologist Friedrich Kützing (1807–1893), most scientists rejected this biological theory of fermentation and accepted instead the chemical theory advocated by Berzelius and by the German chemists Friedrich Wöhler (1800–1882) and Justus von Liebig (1803–1873). In a major paper "Uber die Erischeinungen der Gährung, Fäulniss und Verwesung und ihre Ursachen" (On the Phenomena of Fermentation, Putrefaction and Decay and Their Cause, 1839), Liebig insisted that fermentation had nothing to do with the activity of living organisms but was produced instead by a kind of catalytic chemical reaction.

The controversy over the cause of fermentation was still very much alive in the 1850s, with Liebig's chemical theory remaining dominant. Then, in 1857, there appeared a classic paper, "Memoîre sur la fermentation appelée lactique" (Memoir on the Fermentation Called Lactic), by the great French scientist Louis Pasteur (1822–1895). Pasteur, who in 1854 had been appointed Dean of the Faculty of Sciences at Lille, argued in this paper that lactic acid fermentation was caused by an organized, living creature — a new kind of "yeast," comparable to the ordinary yeast that Schwann and others had claimed to be the cause of alcoholic fermentation. Pasteur showed that lactic acid fermentation was correlated with the development and activity of this new "yeast"—which was later recognized as a species of rod-shaped bacteria, or bacilli (11). When a deposit of these organisms was added to a solution of sugar and chalk, lactic acid fermentation invariably took place. Using similar methods, Pasteur in 1860 demonstrated convincingly that alcoholic fermentation depended upon living yeast cells (12), thus confirming in greater detail and with greater skill the view long held by Schwann. Pasteur later showed that butyric fermentation was also dependent upon living microorganisms.

In the wake of Pasteur's work, only a few scientists still accepted Liebig's original chemical theory of fermentation. Some, however, wondered whether the microorganisms themselves were the active agent in fermentation. Perhaps, they suggested, these microorganisms merely act as the carriers of the real causative agents—enzymes, like Schwann's pepsin and others discovered later. Although several investigators, including Pasteur himself, joined in the search for such fermentative enzymes, they proved elusive. Not until 1897 did the German chemist Edouard Buchner

11 Bacilli of lactic fermentation

(1869–1917) show that fermentation could occur in the absence of living cells. He was able to squeeze out of yeast cells an enzyme, zymase, which could act alone to produce alcoholic fermentation. Subsequent research has confirmed Buchner's general point, though the process is complicated and depends on a variety of subsidiary conditions.

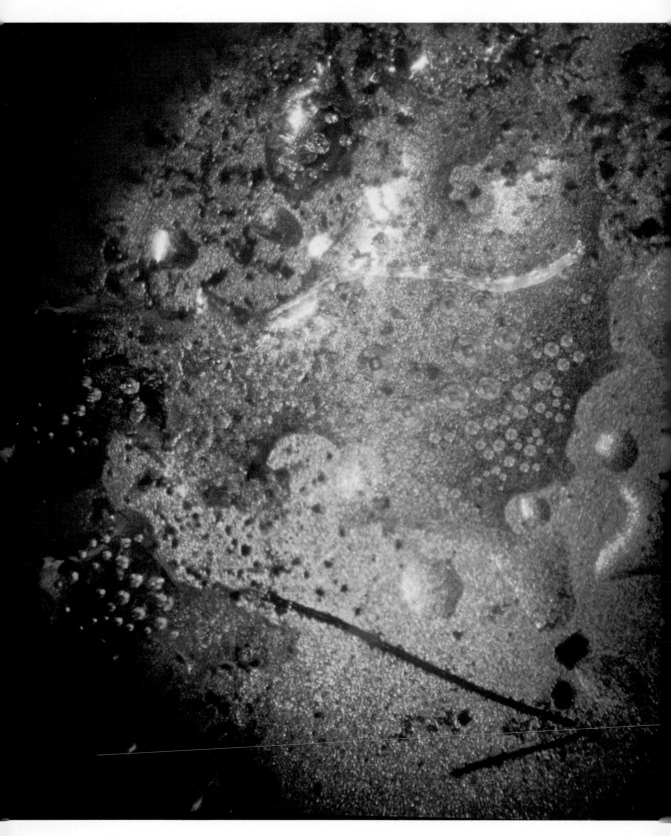

In 1968 a major breakthrough in molecular biology occurred; the enzyme ribonuclease (13) was synthesized. This achievement was made independently by two groups of scientists in the United States—at Rockefeller University by Robert Merrifield and Bernd Gutte, and at Merck & Co. by Robert Denkewalter, Ralph Hirschmann and their associates, who assembled 124 amino acids in two sections and then joined them together to create synthetic RNA-ase.

These accomplishments have tremendous significance. Previously, enzyme synthesis had eluded all investigators, although a *pharmaceutical* method of producing L-asperaginase and urokinase had been discovered. These enzymes are difficult to extract and expensive to produce. Both are life-saving substances. The first breaks up asparagine (14), an amino acid needed by a special kind of leukemia cell. This enzyme is a cure for one type of leukemia. The other enzyme dissolves blood clots; an early injection of it may prevent heart attacks. The discovery of a method of producing these enzymes synthetically could save many lives.

In addition, an understanding of the structure of enzymes could make it possible to create new non-enzyme catalysts that could change the activity of cancer cells directly. A few years ago the synthesis of an enzyme seemed virtually impossible. Suddenly new hopes are rising for significant achievements in curing disease through enzyme synthesis.

12 Fermentation of yeast

13 Ribonuclease, enzyme *(overleaf)*

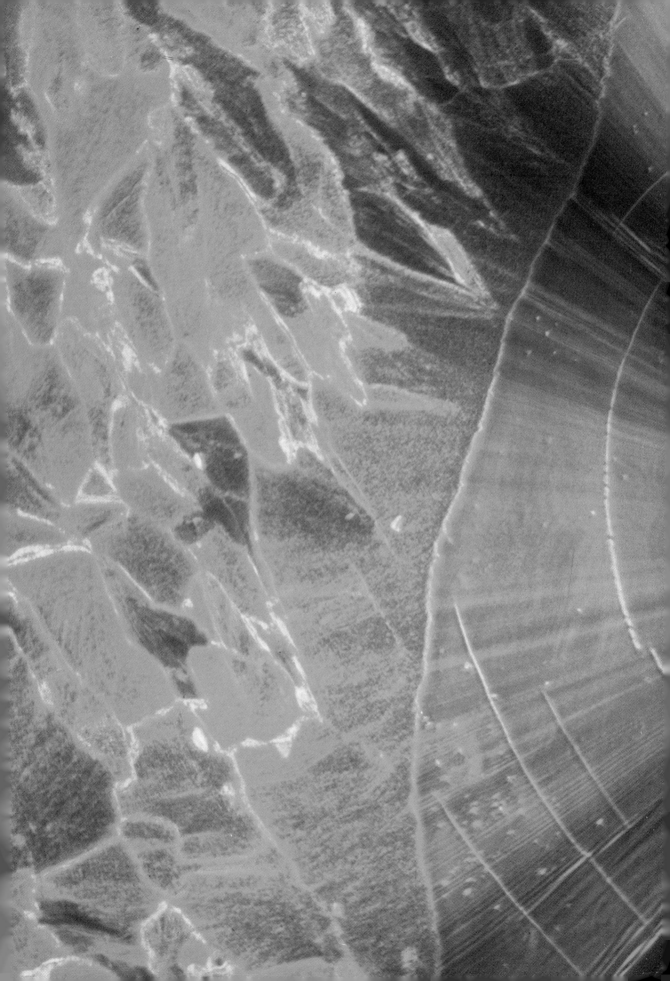

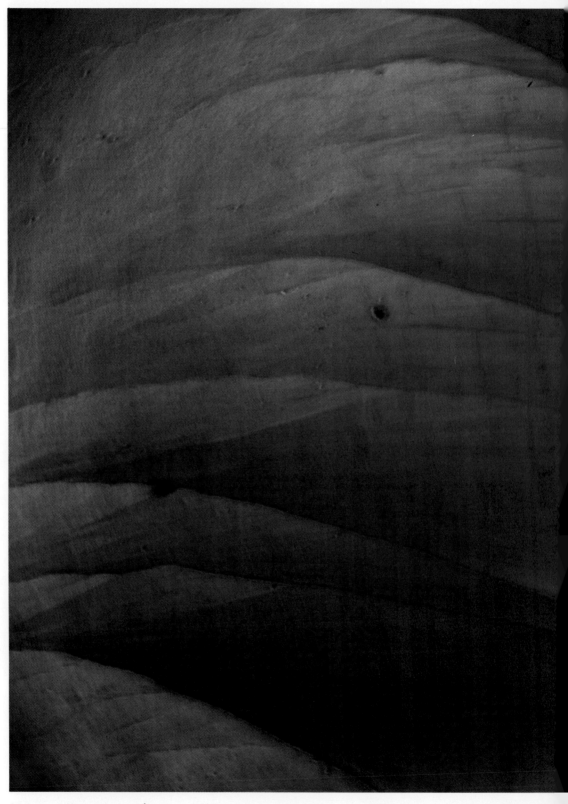

14 Asparagine, amino acid

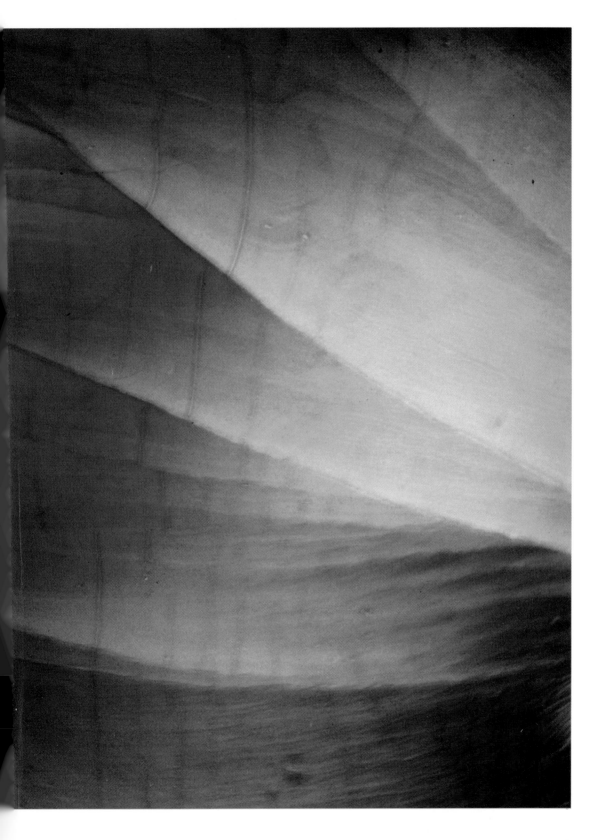

Vitamins

As already mentioned, nine of the amino acids indispensable to man's health are provided by protein in food. Carbohydrates and fats are also necessary, as well as other organic and inorganic materials. Among the necessary organic substances are the vitamins.

What are now called vitamin-deficiency diseases or avitaminoses have been known for centuries. In the age of discovery, voyagers who spent many months on the high seas were often afflicted with the ravages of scurvy. Sailors with the Portuguese explorer Vasco da Gama and the French navigator Jacques Cartier in the sixteenth century were saved by Indians who cured the deficiency attacks with the juice of the Ameda tree (sassafras). Returning mariners reported their experiences, and soon fruit juice became known and widely used as medicine against scurvy. The cause of diseases of this kind, however, remained a mystery until the beginning of the twentieth century.

In a way, the mystery was only deepened by the important developments during the latter half of the nineteenth century in general disease theory and in studies of nutrition. Since disease had long been considered analogous to fermentation, Pasteur's crucial work on fermentation gave new impetus to disease theory as well. Even before Pasteur's work, some theorists had suggested that disease was caused by living organisms. Early in the sixteenth century the Italian physician Girolamo Fracastoro (1483–1553) had supposed that "minute living seeds of disease" caused plague, typhus, and the "French disease"—syphilis, which takes its name from Fracastoro's poem *Syphilis Sine Morbus Gallicius*. By 1840 the German histologist Jacob Henle (1809–1885) was arguing forcefully for the "germ theory of disease." Admitting that he had not proved the theory conclusively, Henle nonetheless established the basic postulates of evidence required to prove the theory. Pasteur's fermentation theory added impressive new support to the germ theory of disease and provided the basis upon which the British surgeon Joseph Lister (1827–1912) built his system of antiseptic surgery. Although the germ theory won an increasing number of supporters, it was not until 1876 that a specific microorganism was established as the cause of a specific disease. In that year Pasteur and the German bacteriologist Robert Koch (1843–1910) showed that rod-shaped bacteria (bacilli) were invariably associated with anthrax, or splenic fever. Intensive study of infective bacteria followed, and a great many of these organisms were isolated between 1880 and 1890, especially by Koch and his associates.

Meanwhile, studies of nutrition had focused on the energy requirements of animals. An elaborate series of investigations, stemming from Liebig's *Animal Chemistry* of 1842 and culminating in the work of the German physiologist Max Rubner (1854–1932), established beyond doubt that the fats, carbohydrates, and proteins supplied in an animal's diet were sufficient for all of its energy needs.

Both of these developments—the germ theory of disease and the energetics approach to nutrition—obscured the significance to health of what came to be called vitamins, which are organic substances required in trace amounts not for their energy value but for their role in maintaining essential metabolic pathways. The story of their discovery begins effectively in 1886, when the Netherlands government sent a group of physicians to the Dutch East Indies to look for the cause of the crippling Oriental disease beriberi. For two and a half years scientists examined tissues of victims under the microscope but could find no bacteria. The discovery of the real cause, like the relatively recent discovery of antibiotics, was stimulated by an accidental observation. Research was being conducted at a penal colony on Java under the guidance of the Dutch physician, Christiaan Eijkman (1883–1930). Chickens were being used as experimental animals, and, for reasons of economy, husks removed in the polishing of rice grains were added to some of their food. Previously sick chickens, unable to fly, walk, or even stand, recovered after being fed these "worthless" husks. Eijkman realized that the outer coats of rice contained a life-restoring substance, but several decades passed before this substance was identified as thiamine, or vitamin B_1.

Earlier, in 1881, the Russian scientist N. Lunin, while working in the laboratory of Gustav von Bunge (1844–1920) in Basel, had found that mice did not survive on what was then regarded as a sufficient diet of proteins, carbohydrates, and fats; they required in addition 2 cubic centimeters of fresh milk daily. Lunin, however, did not pursue the implications of his suggestion that "a natural food such as milk must therefore contain besides [carbohydrates, fats and proteins] small quantities of unknown substances essential to life." Nor was the title of his paper of 1881—"On the significance of inorganic salts for animal nutrition"—likely to draw attention to the possible existence of the organic substances now called vitamins. Vastly more influential was the work of Frederick Gowland Hopkins (1861–1947), professor of biochemistry at Cambridge University, England. From 1907 on, Hopkins emphasized that adequate nutrition depends not only on sufficient calories, protein, and minerals, but also on certain unknown "accessory substances" present in very small amounts. In 1911–1912 Hopkins presented the results of a long series of experiments demonstrating conclusively that rats which failed on a supposedly adequate basal diet were restored to health and vigorous growth by the addition to their diet of small quantities of milk.

In June 1912 the Polish-American biochemist Casimir Funk (1884–1966) published a paper entitled "The Etiology of Deficiency Diseases," in which he suggested that the deficiency diseases beriberi, scurvy, and possibly pellagra and rickets were each caused by the absence from the diet of a specific chemical substance. He had himself, in 1911, already attempted to isolate in pure form the anti-beriberi substance, and despite incomplete success, was confident that it was an organic base or amine. Believing that the absent substances in the other diseases were probably also amines, Funk in 1912 proposed calling all such substances "vitamines" (that is, "life-amines"). It was later found that the anti-beriberi substance (thiamine or vitamin B_1) was the only amine among the accessory food factors, but Funk's term had served as a useful catchword and, with the exception of the final "e," has been used ever since.

Newsworthy events in daily life brought vitamins to the attention of the general public. Lion cubs suffering from rickets in the London Zoo were cured by adding fresh fat (which contains vitamin D) to their diet. It is remarkable how much an attractive name helps the popularization of a new idea.

Vitamins, like antibiotics, have benefited people's health so much that one forgets how recently these blessings were discovered. They are among the most important tools in the medical bag of today's physicians.

15 Vitamin A

There are more than fifteen important vitamins. Vitamin A (15) is found in butter, eggs, and fish oils; a deficiency of it can produce nyctalopia, night blindness, or xerophthalmia, a serious inflammation of the eyes. Vitamin B_1 thiamine (16), the antiberiberi substance, is present in milk, eggs, meats, and vegetables. Other members of the vitamin-B complex are vitamin B_2, riboflavin (17); B_6, pyridoxin (18); and B_{12} (19), cyanocobalamin, deficiency of which produces pernicious anemia. Vitamin C, ascorbic acid (20), occurring in fresh orange and tomato juice, spinach, and some

16 Vitamin B$_1$, thiamine

other vegetables, prevents scurvy and in 1970 received a controversial recommendation from Nobel Laureate Linus Pauling as a weapon against the the common cold. Vitamin D (21) is present in egg yolk, milk, and cod-liver oil; deficiency causes rickets, but, unlike other vitamins, it is harmful in large amounts. Vitamin H, biotin (22),which occurs widely in combination, was isolated in 1935. Biotin deficiency can destroy the brain. Vitamin K (23) assists in blood clotting, preventing excessive bleeding. Deficiency of vitamin P, citrin or hesperidin, weakens the resistance of the capillaries to pressure. In general, vitamin inadequacies cause many insufficiencies of the body and weaken its resistance to diseases of all kinds.

17 Vitamin B_2, riboflavin

18 Vitamin B_6, pyridoxin *(overleaf)*

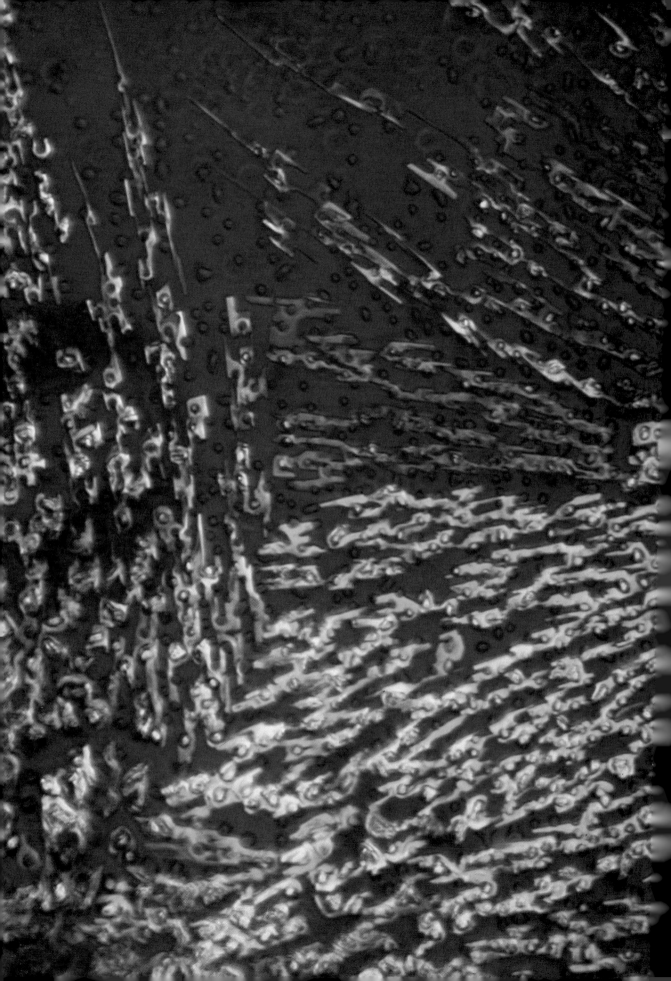

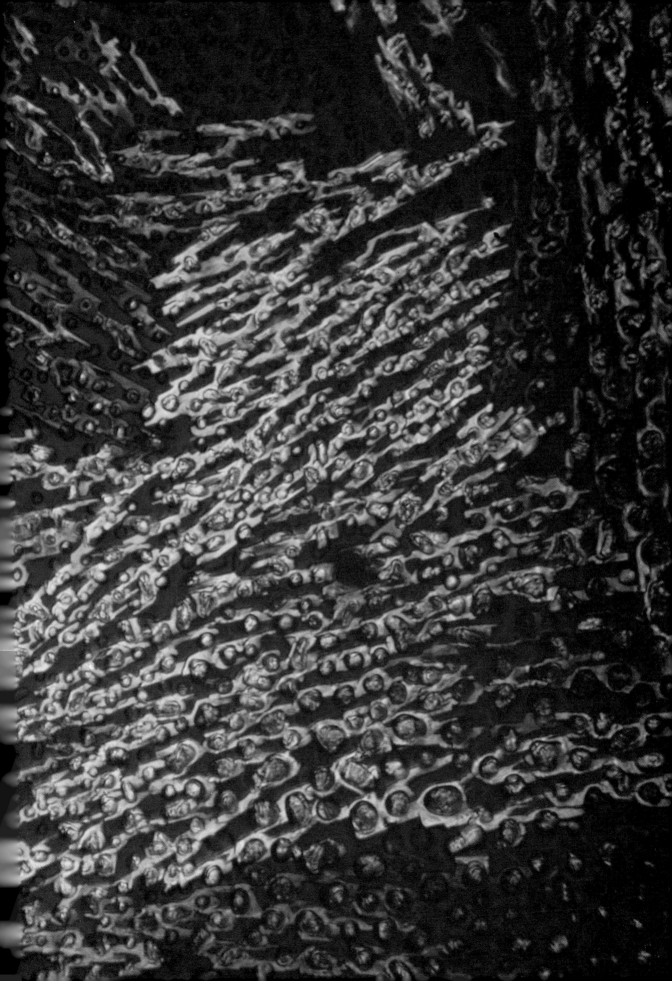

19 Vitamin B_{12}, cyanocobalamin

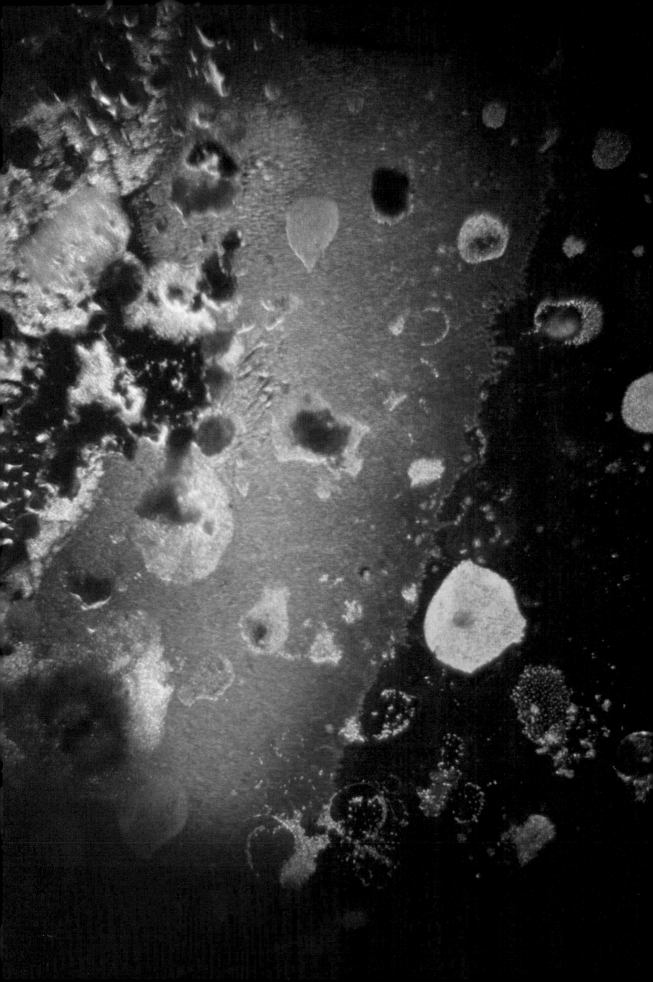

20 Vitamin C, ascorbic acid *(left)*

21 Vitamin D

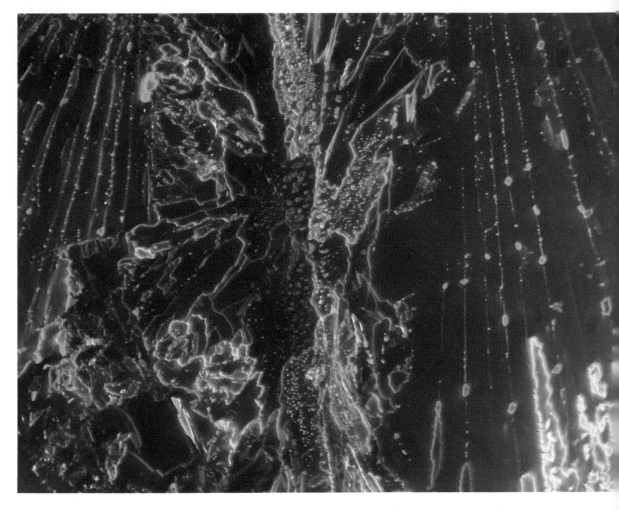

22 Vitamin H, biotin

23 Vitamin K

Hormones

The bodies of animals contain several internal "laboratories" producing complex organic substances. Among these "laboratories" are the salivary, gastric, and intestinal glands, which were known in antiquity. About 1700 the existence of internal glands without ducts leading into other internal organs was assumed. In 1766 the leader of eighteenth-century physiology, the Swiss scientist Albrecht von Haller (1708–1777), discovered the "internal secretion" (so named a century later) of the thyroid and thymus glands, which emit their products directly into the blood. At about the same time, the French physician Théophile de Bordeu (1722–1776) called the attention of Tübingen professors to the fact that castrated males lose their secondary sexual characteristics.

The products of such internal secretions are now called "hormones" (from the Greek, meaning "to arouse"), a term coined in 1905 by the English physiologist E. H. Starling (1866–1927).

Hormones regulate a complex control mechanism, which involves an intricate regulation of metabolic pathways—chemical changes providing energy for fundamental processes in the cell—and vital activities such as assimilation of new materials for regeneration. Hormones are secreted by the pancreas, the adrenal glands, and the pituitary gland of the brain. When a man is frightened, the hormone adrenalin (24) is secreted by glands lying above the kidneys. When adrenalin enters the bloodstream, the heartbeat increases, the blood vessels contract, the blood pressure increases, and the liver releases excess glucose which supplies additional energy.

In 1871 the British clinician Charles Hilton Fagge (d. 1884) associated cretinism with atrophy of the thyroid gland. In 1918 the chemists J. Gudernatsch in Prague and Roman Vishniac in Moscow in separate experiments excised the thyroid glands of tadpoles, which subsequently failed to undergo metamorphosis; feeding with cow's thyroid induced their transformation into frogs. The American biochemist Edward C. Kendall had isolated the active principle, which he named thyroxin (25), 1914.

Diabetes mellitus was associated with pancreas deficiency as early as the eighteenth century, but not until 1921 did the Canadian physicians Frederick Grant Banting (1891–1941) and Charles Herbert Best finally isolate the essential pancreatic secretion, insulin (26), which had been named in 1869. Since 1921 diabetes has been treated successfully with insulin, which controls the burning of carbohydrates. The hormone cortisone (27), isolated by Kendall and others in 1936, and the adrenocorticotropic hormone (ACTH) are beneficial in treating rheumatoid arthritis.

Many hormones, including thyroxin and insulin, are proteins, while others are simpler chemical substances. Continued research is resulting in the discovery of new hormones in human and other mammal bodies. The activity and effects of hormones show a wide variety of functions. Some are involved in general metabolism; others are highly specialized. Hormones are responsible for reproductive processes both in the male—testosterone (28)—and in the female—progesterone (29) as well as for the regulation of pregnancy and the ovaries (prolans), for skeletal growth (phyone), for

24 Adrenalin, hormone

control of gall bladder, for metabolism of sodium, for secretion of insulin, and for many other functions. A hormone of the pituitary gland (30) is responsible for growth in general, while another hormone controls the healing of tissues after injury (31).

Naturally this survey can do no more than provide an introduction to some of life's building blocks, but it has served its purpose if it has shown that proteins, vitamins, and hormones are as essential to life as they are beautiful when viewed through an interference microscope.

25 Thyroxin, hormone *(overleaf)*

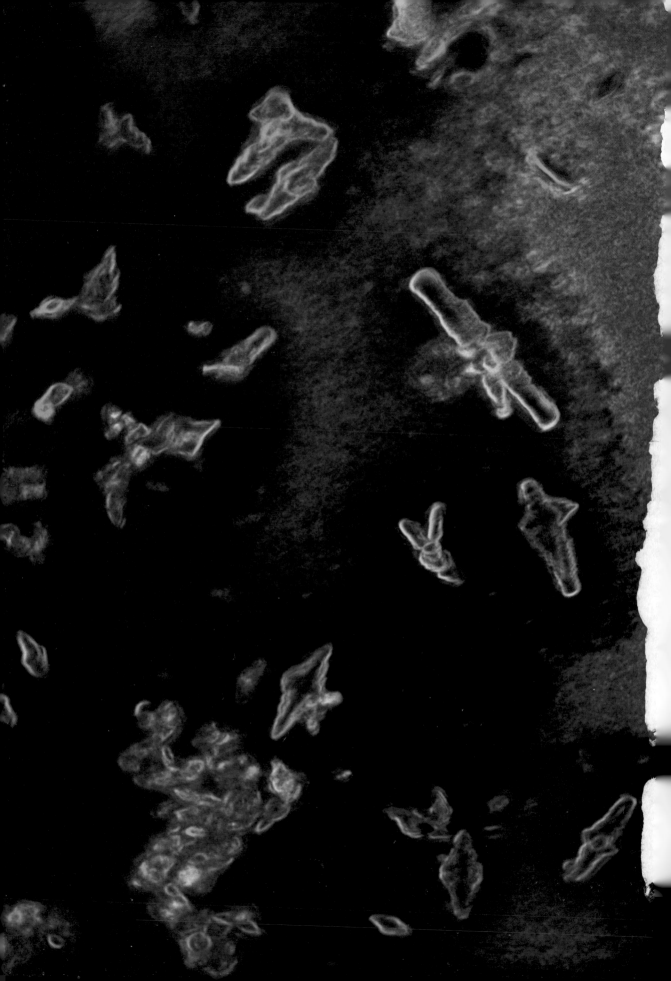

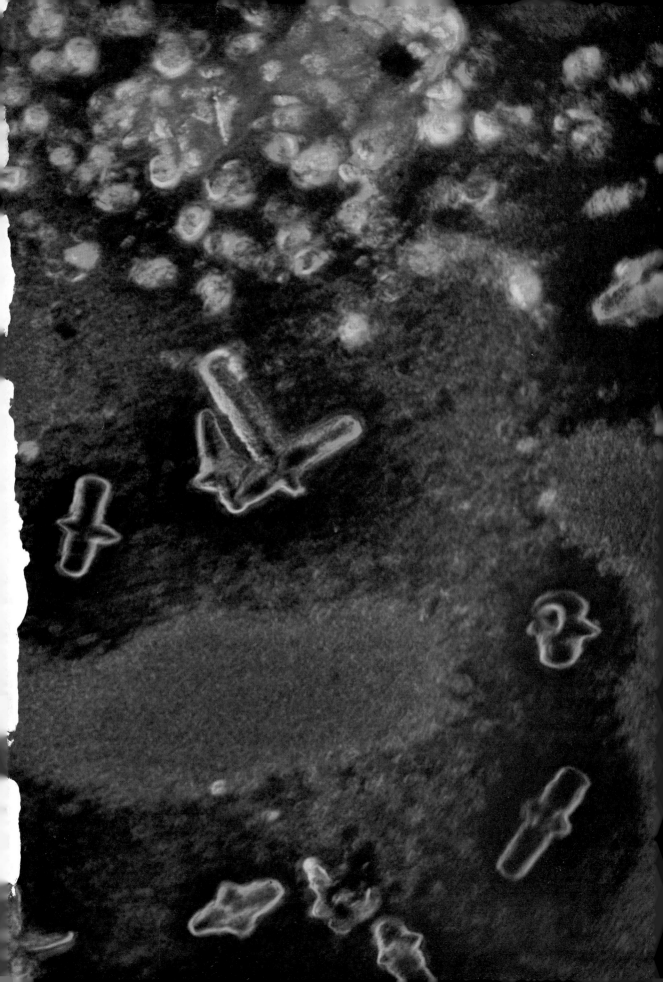

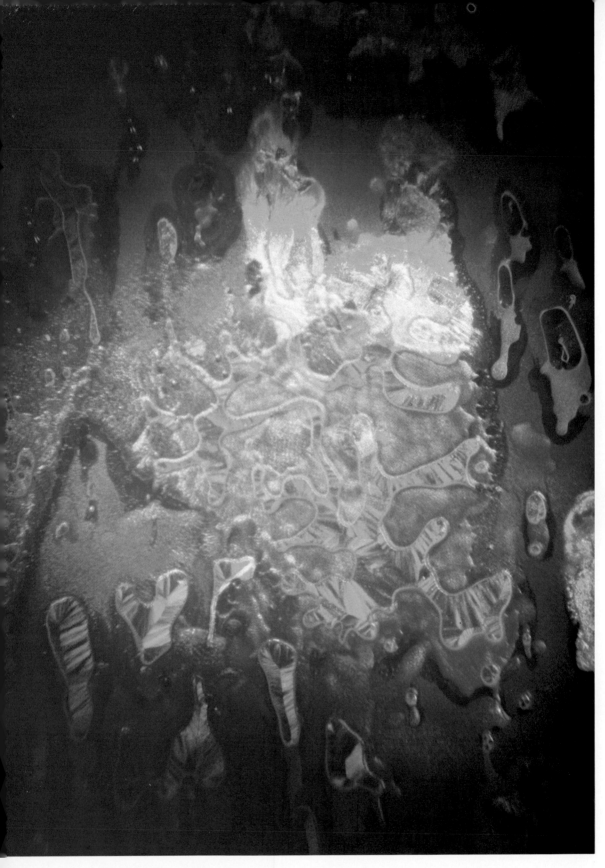

26 Insulin, hormone

27 Cortisone, hormone

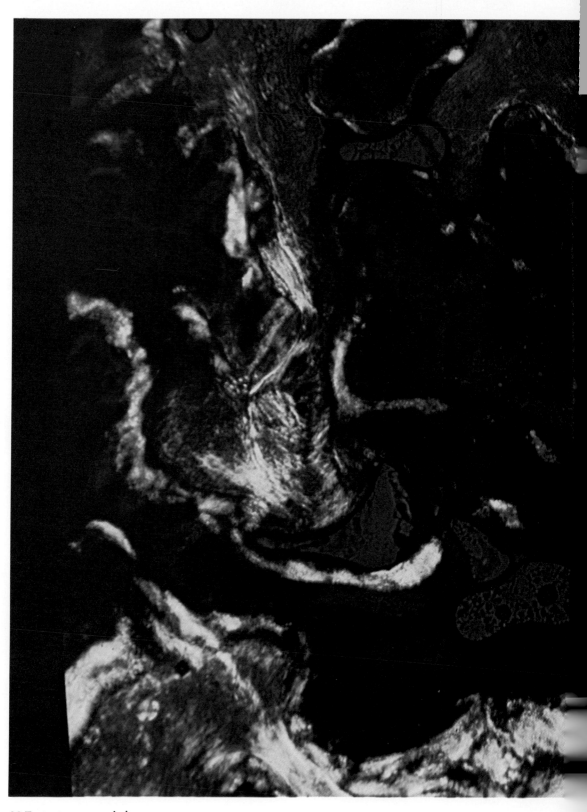

28 Testosterone, male hormone

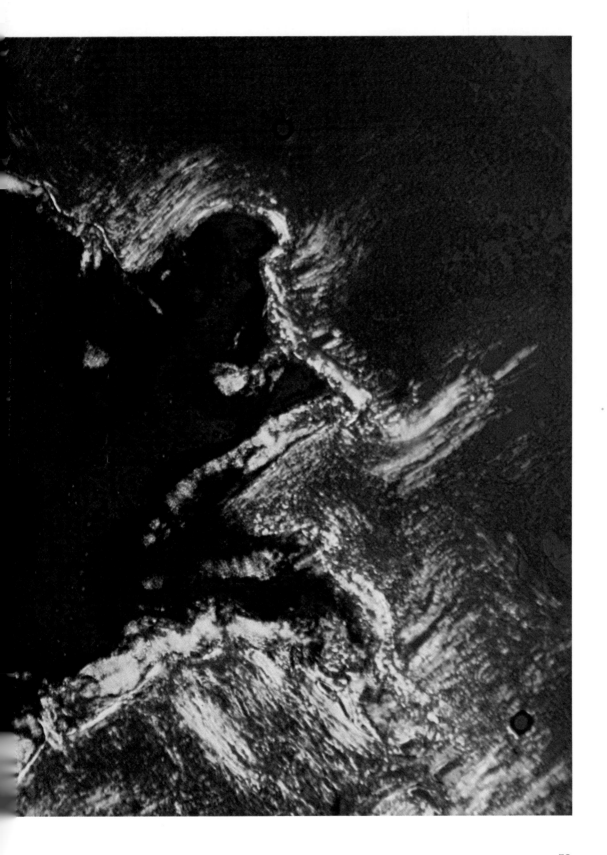

29 Progesterone, female hormone *(overleaf)*

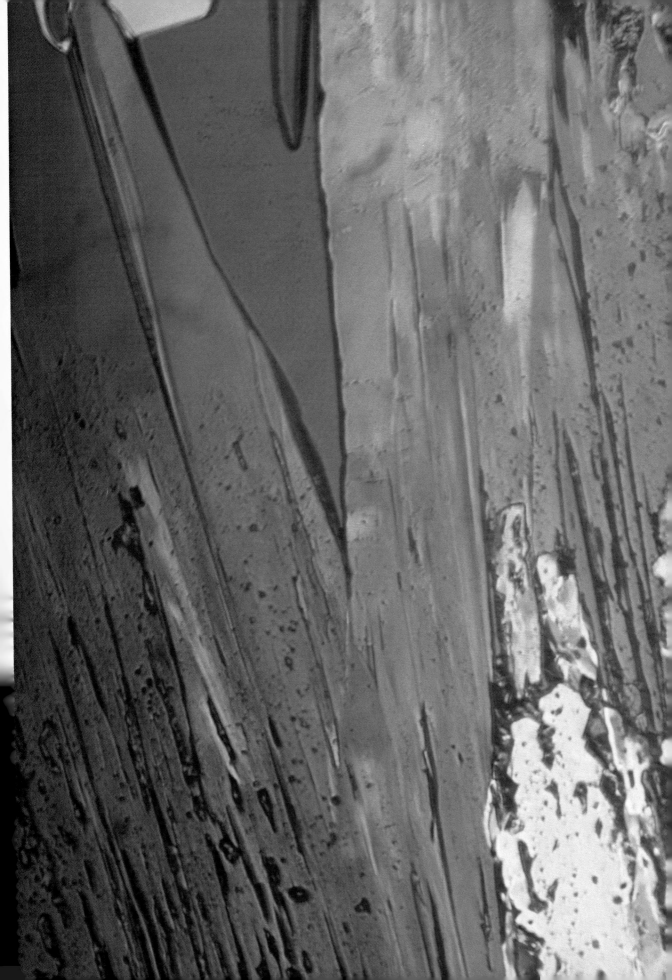

30 Growth hormone of the pituitary gland

31 Tissue-healing hormone

For Further Reading

Beadle, George and Muriel. *The Language of Life*. New York: Doubleday, 1966 (Anchor Book).

Burke, Jack D. *Cell Biology*. Baltimore: Williams & Wilkins, 1970

Clark, B. F. C., and Marcker, K. A. "How Proteins Start." *Scientific American,* January 1968, pp. 36-42.

Handler, Philip (ed.). *Biology and the Future of Man*. New York: Oxford University Press, 1970.

Holley, R. W. "The Nucleotide Sequence of a Nucleic Acid," *Scientific American,* February 1966, pp. 30-39.

Huxley, H. E. "The Mechanism of Muscular Contraction." *Scientific American,* December 1965, pp. 18-27.

Nourse, Alan, and the Editors of *Life. The Body*. New York: Time-Life Books, 1964.

Perutz, Max F. "The Hemoglobin Molecule." *Scientific American,* May 1964, pp. 65-76.

Pfeiffer, John, and the Editors of *Life. The Cell*. New York: Time-Life Books, 1964.

Porter, K. R. "The Ground Substance: Observations from Electron Microscopy." In Brachet, J., and Mirsky, A. E., *The Cell,* Vol. 2. New York: Academic Press, 1961.

Sebrell, William, Jr., Haggerty, James, and the Editors of *Life. Food and Nutrition*. New York: Time-Life Books, 1967.

Thompson, E. O. P. "The Insulin Molecule." *Scientific American,* May 1955, pp. 36-41.

Watson, James D. *The Double Helix*. New York: Atheneum, 1968.

———. *The Molecular Biology of the Gene*. New York: W. A. Benjamin, 1965.

Weiss, P. "The Cell as a Unit." *Journal of Theoretical Biology* 5: 389, 1963.

Index

(Page numbers in italics refer to illustrations)

ACTH (adrenocorticotropic hormone), 46
adrenal glands, 46
adrenalin, 46, *47*
alcohol fermentation, 24, 25
Ameda tree (sassafras), 32
amines, 33
amino acids, 8, 11, 16, 24
anthrax, 32
ascorbic acid (vitamin C), 35, *42*
asparagine, 27, *30-31*
avitaminoses (vitamin-deficiency diseases), 32

bacilli, 24, 32; of lactic fermentation, *25*
Banting, Frederick Grant, 46
Beaumont, William, 18
beriberi, 33
Berzelius, Jöns Jacob, 8, 24
Best, Charles Herbert, 46
biotin (vitamin H), 33, *44*
body tissues, 16
bone-making cells, 16
Bordeu, Théophile de, 46
Braconnet, Henri, 11
Buchner, Edouard, 24-25
Bunge, Gustav von, 33
butyric fermentation, 24

Cagniard-Latour, Charles, 24
cancer cells, 27
carbohydrates, 32, 33
carbon dioxide, 16
Cartier, Jacques, 32
casein, 11
catalytic substances, 16, 27
citrin or hesperidin (vitamin P), 37
"coction," 18
collagen, 16
cortisone, 46, *51*
cretinism, 46
cyanocobalamin (vitamin B_{12}), 35, *40-41*

denaturation process, 16
Denkewalter, Robert, 27
diabetes mellitus, 46
digestion, 18, 20
digestive juices, 24

disease, germ theory of, 32
DNA (deoxyribonucleic acid), *2*, 7
ductless glands, 46

Eijkman, Christiaan, 33
enzymes, 16, 18; and fermentation, 24
enzyme synthesis, 27

Fagge, Charles Hilton, 46
fats, 32, 33
fermentation:
    alcoholic, 24, 25
    butyric, 24
    chemical theory of, 24-25
    and disease, 32
    and enzymes, 24
    and food changes, 18
    and microorganisms, 24-25
    of yeast, *26-27*
Fischer, Emil, 16
Fracastoro, Girolamo, 32
fruit juice, 32
Funk, Casimir, 33

Gama, Vasco da, 32
gastric fistula, 20
gastric juice (of chicken), *18-19*
gastric juice (of human), *20-21*
gelatin, 11
glycine, 8, *10*
growth hormone (of pituitary gland), *56*
Gudernatsch, J., 46
Gutte, Bernd, 27

Haller, Albrecht von, 46
hemoglobin, 16; crystals, *17*
Henle, Jacob, 32
Hirschman, Ralph, 27
histidine, 11, *11*
Hopkins, Frederick Gowland, 33
hormones, 46
hydrolyses, 8, 16

insulin, 46, *50*
"internal secretion," 46

Kendall, Edward C., 46
Koch, Robert, 32
Kützing, Friedrich, 24

lactic acid fermentation, 24
L-asparaginase, 27
leucine, 11, *15*
leukemia, 27
Liebig, Justus von, 24
light waves, interaction of, 7
Lister, Joseph, 32
Lunin, N., 33
lysine, 11, *14*

Merrifield, Robert, 27
milk, 33
Mulder, Gerard Johann, 8

night blindness (nycyalopia), 35
nutrition, 32, 33

optical systems, 7
ovalbumin, 16
oxidation, 16
oxygen, 16

pancreas deficiency, 46
Pasteur, Louis, 24, 32
Pauling, Linus, 37
Pavlov, Ivan P., 20
pellagra, 33
pepsin, 20, *22-23*, 24
phyone, 46
pituitary gland, 46, 47, *56*
polypeptide chains, 16
progesterone, 46, *54-55*
prolans, 46
proteins, 8-16, 32, 33
pyridoxin (vitamin B$_6$), 35, *38-39*

Réaumur, René Antoine, 18
red cells, 16
riboflavin (vitamin B$_2$), 35, *37*
ribonuclease, 27, *28-29*
rickets, 33, 37
Rubner, Max, 32

St. Martin, Alexis, 18, 20
sassafras (Ameda tree), 32
Schwann, Theodor, 20, 24
scurvy, 32, 33
silk fibroin, *8-9*
Starling, E. H., 46
syphilis, 32

testosterone, 46, *52-53*
thiamine (vitamin B$_1$), 33, 35, *36*
thymus gland, 46
thyroid gland, 46
thyroxin, 46, *48-49*
tissue-healing hormones, 57
tryptophan, 11, *12-13*

urokinase, 27

Vishniac, Roman, 46
vitamin-deficiency diseases
    (avitaminoses), 32
vitamins:

A, *34-35*, 35
B$_1$ (thiamine), 33, 35, *36*
B$_2$ (riboflavin), 35, *37*
B$_6$ (pyridoxin), 35, *38-39*
B$_{12}$ (cyanocobalamin), 35, *40-41*
C (ascorbic acid), 35, *42*
D, 34, 37, *43*
H (biotin), 37, *44*
K, 37, *45*
P (citrin or hesperidin), 37

Wöhler, Friedrich, 24

xerophthalmia, 35

yeast, 24; fermentation of, *26-27*

zymase, 25

About the Photographer and Author

Roman Vishniac was born in Russia in 1897 and emigrated to Germany in 1920. He holds a degree in biology and medicine from the University of Moscow and a PH.D. in art from the University of Berlin. He came to the United States in 1940.

At Yeshiva University Dr. Vishniac was research associate from 1957 at the Albert Einstein College of Medicine and professor of biological education from 1961. His chief researches were in the fields of microbiology, particularly in regard to the physiology of ciliates and the study of circulation in unicellular plants.

A pioneer in time-lapse technique in cinematography, Dr. Vishniac is currently associated with the City College of New York, working on a National Science Foundation film-making grant. His photographs are in the permanent collections of many museums, including the Louvre in Paris, the Museum of Modern Art in New York, and the Jerusalem Museum. He has received many awards and citations for his work.

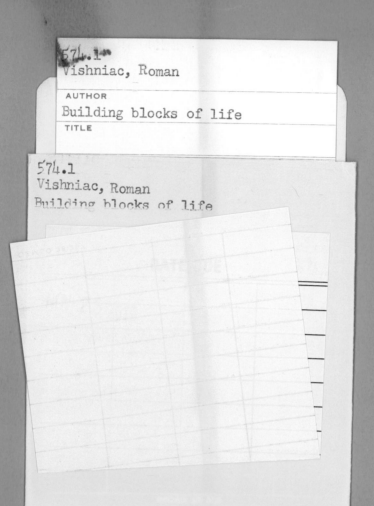

574.1
Vishniac, Roman
AUTHOR
Building blocks of life
TITLE

574.1
Vishniac, Roman
Building blocks of life